Robert L. Linn
University of Colorado

Norman E. Gronlund
University of Illinois

Student Exercise Manual for

Measurement and Assessment in Teaching

7TH EDITION

Merrill,
an imprint of Prentice Hall

Englewood Cliffs, New Jersey **Columbus, Ohio**

Cover art: Paul Klee, *Castle & Sun* (Private collection)
Editor: Kevin M. Davis
Developmental Editor: Carol S. Sykes
Production Buyer: Deidra Schwartz

A Simon & Schuster, Company
Englewood Cliffs, New Jersey 07632

Printed in the United States of America

10 9 8 7 6 5 4 3 2 1

ISBN 0-13-192718-3

Prentice-Hall International (UK) Limited, *London*
Prentice-Hall of Australia Pty. Limited, *Sydney*
Prentice-Hall Canada Inc., *Toronto*
Prentice-Hall Hispanoamericana, S. A., *Mexico*
Prentice-Hall of India Private Limited, , OH *New Delhi*
Prentice-Hall of Japan, Inc., *Tokyo*
Simon & Schuster Asia Pte. Ltd., *Singapore*
Editora Prentice-Hall do Brasil, Ltda., *Rio de Janeiro*

TO THE STUDENT

There are 475 objective test items and more than 100 supply-type items in this manual. Each set of items is based on a clearly stated outcome, to illustrate how test items should be designed to measure specific types of student performance.

These exercises are intended to help you understand and apply the concepts and principles of educational measurement. Although they were especially prepared to use with the seventh edition of Measurement and Assessment in Teaching, efforts were made to emphasize the content and skills that are common to most beginning measurement courses.

The content and organization of this manual are as follows:

- The exercises are arranged by chapter and there are five pages of exercises for each chapter and for the statistics section in Appendix A.

- Each exercise page typically contains a series of selection-type items and supply-type problems, related to specific outcomes they are designed to measure.

- The 25 to 30 selection-type items for a given chapter enable students to check their knowledge and understanding of basic concepts. The answers on the back page of each exercise provide immediate feedback.

- The supply-type problems for each chapter provide students with an opportunity to test their understanding further and to apply what they are learning to practical situations (e.g., constructing test items, developing performance-based assessment tasks and scoring guides, interpreting item analysis data, test scores, and other statistics). Where computation is involved, the final answer is provided on the back of the answer sheet.

Studying the textbook and doing the exercises in the manual should provide a sound foundation for using measurement and assessment in teaching.

CONTENTS

CONTENTS

CONTENTS

Appendix A - Elementary Statistics

Exercise 1-A

PRINCIPLES AND PROCEDURES OF CLASSROOM ASSESSMENT

OUTCOME: Distinguishes between sound and unsound principles and procedures.

Directions: Indicate whether each of the following statements represents a **sound (S)** or **unsound (U)** principle or procedure of classroom assessment by circling the appropriate letter to the left of the statement.

S U 1. The first step in measuring classroom learning is to decide on the type of test to use.

S U 2. Classroom assessment should be based on objective data only.

S U 3. The type of classroom assessment used should be determined by the performance to be measured.

S U 4. Effective classroom assessment requires the use of a variety of assessment techniques.

S U 5. Assessment techniques should replace teacher observation and judgment.

S U 6. Error of measurement must always be considered during the interpretation of assessment results.

OUTCOME: State the meaning of test, measurement, and assessment.

Directions: In your own words, state the meaning of each of the following terms.

Test:

Measurement:

Assessment:

(Turn up bottom of page for answers)

Answers to SELECTION (1-A)

1.U 2.U 3.S

4.S 5.U 6.S

Exercise 1-B

CLASSROOM ASSESSMENT AND THE INSTRUCTIONAL PROCESS

OUTCOME: Identifies how classroom assessment functions in the instructional process.

Directions: Indicate whether the textbook author would **agree (A)** or **disagree (D)** with each of the following statements by circling the letter to the left of the statement.

A D 1. The main purpose of classroom assessment is to improve learning.

A D 2. The first step in both teaching and assessment is to determine the intended learning outcomes.

A D 3. Classroom assessments should not be given until the end of instruction.

A D 4. Instructional objectives should aid in selecting the types of assessment instruments to use.

A D 5. Assessment results should be used primarily for assigning grades.

A D 6. Classroom assessment always includes some form of testing.

OUTCOME: Describes the role of instructional objectives.

Directions: Describe the role of instructional objectives in classroom assessment.

(Turn up bottom of page for answers)

Answers to SELECTION (1-B)

1.A 2.A 3.D

4.A 5.D 6.D

Exercise 1-C

MEANING OF PLACEMENT, FORMATIVE, DIAGNOSTIC, AND SUMMATIVE ASSESSMENT

OUTCOME: Classifies examples of classroom assessment procedures.

Directions: For each of the following descriptions, indicate which **type of assessment** is represented by circling the appropriate letter using the following key.

KEY P = Placement F = Formative
D = Diagnostic S = Summative

P F D S 1. An achievement test is used to certify student mastery.

P F D S 2. Students are given a ten-item test to determine learning progress.

P F D S 3. A teacher observes the process used as a student solves arithmetic problems.

P F D S 4. Algebra students take an arithmetic test on the first day of class.

P F D S 5. Course grades are assigned.

P F D S 6. An assessment is given at the beginning of a new unit.

OUTCOME: State examples of types of assessment procedures.

Directions: For each of the following types of assessment state one specific example that illustrates its use in some subject area.

Placement:

Formative:

Diagnostic:

Summative:

(Turn up bottom of page for answers)

Answers to SELECTION (1-C)

1.S 2.F 3.D

4.P 5.S 6.P

Exercise 1-D

MEANING OF CRITERION-REFERENCED AND NORM-REFERENCED INTERPRETATIONS

OUTCOME: Distinguishes between examples of each type of interpretation.

Directions: Indicate whether each of the following statements represents a **criterion-referenced (C)** interpretation or a **norm-referenced (N)** interpretation by circling the appropriate letter.

C N 1. Erik obtained the highest score on the reading test.

C N 2. Carlos can identify all of the parts of a sentence.

C N 3. Connie can type 60 words per minute.

C N 4. John earned an average score on an arithmetic test.

C N 5. Tonia defined only 20 percent of the science terms.

C N 6. Maria set up her laboratory equipment faster than anyone else.

OUTCOME: Writes statements representing each type of interpretation.

Directions: Write three statements that represent **criterion-referenced** interpretations and three statements that represent **norm-referenced** interpretations.

Criterion-referenced interpretations:

Norm-reference interpretations:

(Turn up bottom of page for answers)

Answers to SELECTION (1-D)

1.N 2.C 3.C

4.N 5.C 6.N

Exercise 1-E

MEANING OF CONTRASTING TEST TYPES

OUTCOME: Distinguishes between contrasting test types.

Directions: For each of the following test descriptions indicate which **test type** is represented by circling the letter to the left of each description using the following key.

KEY					
	A=Informal	C=Mastery	E=Speed	G=Objective	I=Verbal
	B=Standardized	D=Survey	F=Power	H=Subjective	J=Performance

A B 1. A test using national norms for interpretation.

C D 2. A test used to measure many skills with just a few items for each skill.

E F 3. A test that all students are expected to complete.

G H 4. A test on which different scorers obtain the same results.

I J 5. A test requiring students to describe how to set up laboratory equipment.

OUTCOME: Describes a test representing a given test type.

Directions: In the spaces below, write a brief description of a specific test representing each of the test types.

Survey test:

Mastery test:

Power test:

Objective test:

(Turn up bottom of page for answers)

Answers to SELECTION (1-E)

1.B 2.D 3.F 4.G 5.I

Exercise 2-A

INSTRUCTIONAL OBJECTIVES AS LEARNING OUTCOMES

OUTCOME: Distinguishes between statements of learning process and learning outcomes.

Directions: Indicate whether each of the following features describes a **learning process (P)** or a **learning outcome (O)**.

P O 1. Learns the meaning of terms.

P O 2. Develops a more favorable attitude toward reading.

P O 3. Demonstrates concern for the environment.

P O 4. Locates a position on a map.

P O 5. Practices interpreting charts and graphs.

P O 6. Describes the value of good study habits.

OUTCOME: Writes well-stated outcomes.

Directions: In the spaces below (1) rewrite as learning **outcomes** each of the statements at the top of the page that were classified as learning **processes**, and (2) write three general statements of learning **outcomes** for a course of subject area.

1. Learning outcomes rewritten from process statements.

2. Three general learning outcomes for a course.

(Turn up bottom of page for answers)

Answers to SELECTION (2-A)

1.P   2.P   3.O

4.O   5.P   6.O

Exercise 2-B

DOMAINS OF THE TAXONOMY (COGNITIVE, AFFECTIVE, PSYCHOMOTOR)

OUTCOME: Identifies examples of instructional objectives belonging to each **Taxonomy**.

Directions: Indicate the **Taxonomy** domain to which each of the following general instructional objectives belongs by circling the appropriate letter.

KEY A = Affective
C = Cognitive
P = Psychomotor

A C P 1. Understands basic concepts.

A C P 2. Appreciates the contributions of scientists.

A C P 3. Evaluates a book.

A C P 4. Operates a slide projector.

A C P 5. Writes smoothly and legibly.

A C P 6. Demonstrates an interest in science.

OUTCOME: Writes general instructional objectives that fit each **Taxonomy** domain.

Directions: Write two general instructional objectives for each of the following domains of the **Taxonomy**.

Cognitive objectives:

Affective objectives:

Psychomotor objectives:

(Turn up bottom of page for answers)

Answers to SELECTION (2-B)

1.C 2.A 3.C

4.P 5.P 6.A

Exercise 2-C

SELECTING APPROPRIATE INSTRUCTIONAL OBJECTIVES

OUTCOME: Distinguishes between sound and unsound criteria for selecting instructional objectives.

Directions: Indicate whether each of the following statements is a **sound (S)** or **unsound (U)** criterion for selecting instructional objectives by circling the appropriate letter.

S U 1. Instructional objectives should be limited to those learning outcomes that can be measured objectively.

S U 2. Instructional objectives should be related to the goals of the school.

S U 3. Instructional objectives should be concerned primarily with knowledge of facts.

S U 4. Instructional objectives should be selected in terms of their feasibility.

S U 5. Instructional objectives should specify the intended learning outcomes.

OUTCOME: Describes the importance of selecting appropriate instructional objectives.

Directions: In your own words, describe the importance of carefully selecting instructional objectives.

(Turn up bottom of page for answers)

Answers to SELECTION (2-C)

1.U 2.S 3.U 4.S 5.S

Exercise 2-D

STATING GENERAL INSTRUCTIONAL OBJECTIVES

OUTCOME: Distinguishes between well-stated and poorly stated general instructional objectives.

Directions: For each of the following pairs of objectives, indicate the one that is **best stated** as a general instructional objective by circling the letter of your answer (A or B).

1. A Reads supplementary references.
   B Sees the importance of reading.

2. A Is aware of the value of money.
   B Comprehends oral directions.

3. A Shows students how to make accurate computations.
   B Judges the adequacy of an experiment.

4. A Demonstrates proficiency in laboratory skills.
   B Gains minimum proficiency in mathematics.

5. A Studies weather maps.
   B Constructs weather maps.

6. A Is familiar with the use of the library.
   B Locates references in the library.

OUTCOME: Rewrites poorly stated objectives as **well-stated** general instructional objectives.

Directions: Rewrite as **well-stated** general instructional objectives each of the six poorly stated objectives at the top of the page.

1.

2.

3.

4.

5.

6.

(Turn up bottom of page for answers)

Answers to SELECTION (2-D)

1.A 2.B 3.B

4.A 5.B 6.B

Exercise 2-E

STATING SPECIFIC LEARNING OUTCOMES

OUTCOME: Distinguishes between performance and non-performance statements of specific learning outcomes.

Directions: For each of the following pairs of specific learning outcomes, indicate the one that is stated in **performance terms**.

1. A States the principle.
   B Realizes the value of the principle.

2. A Increases ability to read.
   B Selects the main thought in a passage.

3. A Learns facts about current events.
   B Relates facts in explaining current events.

4. A Distinguishes facts from opinions.
   B Is aware that opinions should not be stated as facts.

5. A Grasps the meaning of terms when used in context.
   B Defines the terms in his or her own words.

6. A Identifies the value of a given point on a graph.
   B Determines the trend shown in a graph.

OUTCOME: States specific learning outcomes in performance terms.

Directions: Write three specific learning outcomes, in **performance terms**, for each of the following general instructional objectives.

Knows basic terms.

Demonstrates good study habits.

Interprets a weather map.

(Turn up bottom of page for answers)

Answers to SELECTION (2-E)

1.A 2.B 3.B

4.A 5.B 6.A

Exercise 3-A

VALIDITY AND RELATED CONCEPTS

OUTCOME: Identifies the nature of validity.

Directions: Indicate which of the following statements concerning validity are **correct (C)** and which are **incorrect (I)** by circling the appropriate letter.

C I 1. Validity refers to the test itself, not just the test scores.

C I 2. Validity is a matter of degree (e.g., high, low).

C I 3. Validity is a general quality that applies to various uses of assessment results.

C I 4. Validity is a unitary concept.

C I 5. Validity refers to how consistently a test measures.

C I 6. Validity is adequately described by the correlation of assessment scores with a criterion measure.

OUTCOME: Distinguishes among **validity**, **reliability**, and **usability**.

Directions: Briefly describe the key feature of each concept.

Validity:

Reliability:

Usability:

(Turn up bottom of page for answers)

Answers to SELECTION (3-A)

1.I 2.C 3.I

4.C 5.I 6.I

Exercise 3-B

MAJOR VALIDITY CONSIDERATIONS

OUTCOME: Identifies characteristics of the major validity considerations.

Directions: For each of the following statements, indicate which **major validity consideration** is being described by circling the appropriate letter using the following key.

KEY: A = content B = test-criterion relationships
C = construct D = consequences

A B C D 1. Can be expressed by an expectancy table.

A B C D 2. Is determined primarily by logical analysis.

A B C D 3. Evaluates what happens when assessment results are used.

A B C D 4. Typically requires more than one criterion.

A B C D 5. Emphasizes the representativeness of the sample of tasks.

A B C D 6. The results may be summarized by a single number.

OUTCOME: Writes an example that illustrates each of the four major validity considerations.

Directions: Briefly describe an example of evidence that would be relevant for each major consideration.

Content:

Test-Criterion Relationships:

Construct:

Consequences:

(Turn up bottom of page for answers)

Answers to SELECTION (3-B)

1.B 2.A 3.D

4.C 5.A 6.B

Exercise 3-C

MEANING OF CORRELATION

OUTCOME: Interprets correlation coefficients and the effects of various conditions on them.

Directions: In each of the following pairs of statements, select the statement that indicates the **greater degree of relationship** and circle the letter of your answer (A or B). Assume other things are equal.

1. A A correlation coefficient of .60.
   B A correlation coefficient of .10.

2. A A predictive validity coefficient.
   B A concurrent validity coefficient.

3. A A predictive validity coefficient of .70.
   B A predictive validity coefficient of -.80.

4. A A correlation between test scores and a criterion measure obtained one week later.
   B A correlation between test scores and a criterion measure obtained one year later.

5. A The predictive validity coefficient for a set of scores ranging from 20 to 100.
   B The predictive validity coefficient for a set of scores ranging from 80 to 100.

OUTCOME: Lists factors influencing a correlation coefficient.

Directions: List three factors that will cause correlation coefficients to be small.

(Turn up bottom of page for answers)

Answers to SELECTION (3-C)

1.A 2.B 3.B 4.A 5.A

Exercise 3-D

EXPECTANCY TABLE

OUTCOME: Interprets an expectancy table.

Directions: In the expectancy table below, the row for each score level shows the **percentage** of students who earned a grade of A, B, C, D, or F. Review the table and answer the question following it ("Chances" means chances in 100).

	Percentage of Students					
Score	F	D	C	B	A	Total
115 - 134	0	12	20	26	42	100
95 - 114	10	18	18	24	30	100
75 - 94	32	26	18	18	6	100

_______ 1. If Sara had a score of 120, what are her chances of obtaining a grade of A?

_______ 2. If Bob had a score of 113, what are his chances of obtaining a failing grade?

_______ 3. If Tanya had a score of 90, what are her chances of obtaining a grade of C or higher?

_______ 4. How many times greater are Sara's chances than Tanya's of obtaining a grade of A?

_______ 5. What score levels provides the **best** prediction?

_______ 6. What score levels provides the **poorest** prediction?

OUTCOME: Describes the advantages, limitations, and cautions in using expectancy tables.

Directions: In the appropriate spaces below, describe the advantages, limitations, and cautions in using expectancy tables.

Advantages:

Limitations:

Cautions in Interpreting:

(Turn up bottom of page for answers)

Answers to SELECTION (3-D)

1. 42	2. 10	3. 42
4. 7	5. 115-134	6. 95-114

Exercise 3-E

FACTORS AND CONDITIONS INFLUENCING VALIDITY

OUTCOME: Identifies the influence of assessment practices on validity.

Directions: Indicate what **influence** each of the following assessment practices is most likely to have on validity by circling the appropriate letter to the left of each statement, using the following key.

KEY R = Raise validity L = Lower validity

R L 1. Increase item difficulty by using more complex sentence structure.

R L 2. Increase the number of items measuring each specific skill from five to ten.

R L 3. Replace multiple-choice items with short-answer items for measuring the ability to define terms.

R L 4. Replace multiple-choice items by laboratory performance tasks for measuring ability to conduct experiments.

R L 5. Use selection-type items instead of supply-type items to measure spelling ability.

R L 6. Use an essay test to measure factual knowledge of historical events.

OUTCOME: Lists factors that lower assessment validity.

Directions: In the space provided below, list as many factors as you can think of that might lower the validity of a classroom assessment.

(Turn up bottom of page for answers)

Answers to SELECTION (3-E)

1.L 2.R 3.R

4.R 5.L 6.L

Exercise 4-A

COMPARISON OF VALIDITY AND RELIABILITY

OUTCOME: Identifies similarities and differences between validity and reliability.

Directions: Indicate whether each of the following statements is characteristic of **validity (V)**, **reliability (R)**, or **both (B)**, by circling the appropriate letter to the left of the statement.

V R B 1. Can be expressed by an expectancy table or regression equation.

V R B 2. Refers to the consistency of a measurement.

V R B 3. Is often based on a comparison with an external criterion.

V R B 4. Refers to the results of a test rather than the test itself.

V R B 5. Compares performance on two halves of an assessment.

V R B 6. Contributes to more effective classroom teaching.

OUTCOME: Explains the relationship between validity and reliability.

Directions: In the appropriate spaces below, briefly explain each of the following statements.

1. If assessment results are highly valid, they will also be highly reliable.

2. If assessment results are highly reliable, they may or may not be valid.

3. In selecting an assessment, validity has priority over reliability.

(Turn up bottom of page for answers)

Answers to SELECTION (4-A)

1.V 2.R 3.V

4.B 5.R 6.B

Exercise 4-B

METHODS FOR DETERMINING RELIABILITY

OUTCOME: Distinguishes among the methods for determining reliability.

Directions: For each of the following statements, indicate **which method** of determining reliability is being described by circling the appropriate letter using the following key.

KEY: A = Test-retest, same form B = Equivalent form, immediate
C = Equivalent form, time interval D = Split-half

A B C D 1. Provides an inflated reliability coefficient for a speeded test.

A B C D 2. Usually provides the smallest reliability for a given assessment.

A B C D 3. Is best for determining the stability of assessment results.

A B C D 4. Does not indicate consistency of performance over different samples of items.

A B C D 5. Correlation coefficient must be adjusted with the Spearman-Brown formula.

A B C D 6. Best indicator of the adequacy of the test sample.

OUTCOME: Summarizes the procedure for obtaining various types of reliability coefficients.

Directions: Briefly describe the **procedure for obtaining** each type of reliability coefficient.

Test-retest:

Equivalent forms:

Split-half:

Inter-rater agreement:

(Turn up bottom of page for answers)

Answers to SELECTION (4-B)

1.D 2.C 3.A

4.A 5.D 6.B

Exercise 4-C

RELATING RELIABILITY TO THE USE OF ASSESSMENT RESULTS

OUTCOME: Selects the reliability method that is most relevant to a particular use of assessment results.

Directions: For each of the following objectives, select the reliability method that is **most relevant** by circling the appropriate letter using the following key.

KEY: T = Test-retest (time interval) E = Equivalent form (immediate)
S = Split-half I = inter-rater consistency

T E S I 1. Determining whether test scores on school records are still dependable.

T E S I 2. Selecting an achievement test to measure growth over one school year.

T E S I 3. Selecting a highly speeded test.

T E S I 4. Evaluating the adequacy of judgmental scoring of performances on a complex task.

T E S I 5. Seeking support for the adequacy of the sample of test items.

T E S I 6. Determining whether an informal classroom assessment has internal consistency.

OUTCOME: Justifies the selection of a reliability method for a particular test use.

Directions: For each of statements 1-6 above, write a sentence or two to **justify** why you think the selected reliability method would provide the most relevant information for that particular use.

1.

2.

3.

4.

5.

6.

(Turn up bottom of page for answers)

Answers to SELECTION (4-C)

1.T   2.E   3.E

4.I   5.E   6.S

Exercise 4-D

RELIABILITY COEFFICIENT AND STANDARD ERROR OF MEASUREMENT

OUTCOME: Identifies the similarities and differences between the two basic methods of expressing reliability.

Directions: Indicate whether each of the following statements is more characteristic of the **reliability coefficient (R)**, the **standard error of measurement (E)**, or **both (B)** by circling the appropriate letter to the left of the statement.

R E B 1. Indicates the degree to which a set of scores contains error.

R E B 2. Remains relatively constant from one group to another.

R E B 3. Cannot be computed without the other.

R E B 4. Useful in selecting a test for a particular grade.

R E B 5. Increases as the spread of scores increases.

R E B 6. Would be zero if the test were perfectly reliable.

OUTCOME: Describes the use of standard error in interpreting test scores.

Directions: In the appropriate spaces below describe how confidence bands (error bands) are used to interpret each of the following.

1. An individual test score.

2. The difference between two test scores.

(Turn up bottom of page for answers)

Answers to SELECTION (4-D)

1.B 2.E 3.E

4.B 5.R 6.E

Exercise 4-E

FACTORS INFLUENCING RELIABILITY AND INTER-RATER CONSISTENCY

OUTCOME: Identifies the influence of assessment practices on reliability.

Directions: Indicates whether each of the following practices is most likely to **raise (R)**, **lower (L)**, or have **no effect (N)** on reliability.

R L N 1. Add more items like those in the test.

R L N 2. Remove ambiguous tasks from the assessment.

R L N 3. Add five items that everyone answers correctly.

R L N 4. Replace a multiple-choice test with an essay test.

R L N 5. Modify the assessment tasks to obtain a wide spread of scores.

R L N 6. Replace a 10-item multiple-choice quiz with a 10 item true-false quiz.

OUTCOME: Computes and interprets inter-rater consistency expressed as the percent of exact agreement.

Directions: Using the information from the following table compute the percent exact agreement for the scores provided by two independent raters.

		Scores Assigned by Rater 1				
	Score	1	2	3	4	Row Total
Scores Assigned by Rater 2	4	0	1	4	12	17
	3	1	5	17	6	29
	2	6	18	7	1	32
	1	16	5	1	0	22
	Column Total	23	29	29	19	100

Percent exact agreement =

Briefly interpret the results.

(Turn up bottom of page for answers)

Answers to SELECTION (4-E)

1.R 2.R 3.N

4.L 5.R 6.L 63%

Exercise 5-A

TYPES AND USES OF CLASSROOM TESTS AND ASSESSMENTS

OUTCOME: Relates type of test item or assessment task to information needed.

Directions: For each of the following questions, indicate which type of test item or assessment task would provide the **most useful information** by circling the appropriate letter to the left of the question.

KEY: P = Placement F = Formative S = Summative

P F S 1. Are students making satisfactory progress in learning to make connections among major mathematical concepts?

P F S 2. What types of errors are student making in learning grammar?

P F S 3. Should Carman be encouraged to enroll in an advanced mathematics course?

P F S 4. Is Michael ready for instruction on the new unit?

P F S 5. What final grade should Lizanne receive in the science course?

P F S 6. How do my students rank in achievement?

OUTCOME: States whether a criterion-referenced or norm-referenced test is more useful for a particular use and justifies the choice.

Directions: For each of the questions 1-6 above, (1) state whether a **criterion-referenced test** or a **norm-referenced test** would provide the more useful information, and (2) explain, in a sentence or two, why you think that test type would be more useful.

1.

2.

3.

4.

5.

6.

(Turn up bottom of page for answers)

Answers to SELECTION (5-A)

1.F   2.F   3.P

4.P   5.S   6.S

Exercise 5-B

SPECIFICATIONS FOR CLASSROOM TESTS AND ASSESSMENTS

OUTCOME: Identifies the procedures involved in preparing specifications for classroom tests and assessments.

Directions: For each of the following statements, indicate whether the procedure is a **desirable (D)** or **undesirable (U)** practice when preparing specifications for test and assessments by circling the appropriate letter to the left of the statement.

D U 1. Start by identifying the intended learning outcomes.

D U 2. Limit the specifications to those outcomes that can be measured objectively.

D U 3. Consider the instructional emphasis when specifying the sample of items and tasks.

D U 4. Increase the relative weighting of topics by including more items on those topics.

D U 5. Use a table of specifications for summative tests only.

D U 6. Consider the purpose of testing when determining item difficulty.

OUTCOME: Explains the importance and nature of using a table of specifications.

Directions: Briefly explain each of the following statements in the space that follows it.

1. Well-defined specifications contribute to validity.

2. Well-defined specifications contribute to interpretability of the results.

3. Tables of specifications may differ for end of unit and end of course assessments.

(Turn up bottom of page for answers)

Answers to SELECTION (5-B)

1.D 2.U 3.D

4.D 5.U 6.D

Exercise 5-C

USE OF OBJECTIVE ITEMS AND PERFORMANCE ASSESSMENT TASKS

OUTCOME: Identifies whether objective items or performance assessment tasks are more appropriate for a given condition.

Directions: For each of the following conditions, indicate whether **objective (O)** items or **performance (P)** assessment tasks would be more appropriate by circling the correct letter to the left of each statement.

O P 1. A broad sampling of learning outcomes is desired.

O P 2. The need is to measure ability to organize.

O P 3. Time available to prepare test or assessment is short.

O P 4. Time available for scoring is short.

O P 5. The need is for measuring knowledge of important facts and major concepts covered throughout the semester.

O P 6. The need is to measure learning at the synthesis level.

OUTCOME: States whether objective items or performance tasks are more useful for measuring a particular instructional objective and justifies the choice.

Directions: For each of the following general instructional objectives, (1) state whether **objective** items or **performance** tasks would be more appropriate, and (2) explain, in a sentence or two, why you think that approach would be more appropriate.

1. Knows specific facts.

2. Interprets a weather map.

3. Evaluates a plan for an experiment.

(Turn up bottom of page for answers)

Answers to SELECTION (5-C)

1.O 2.P 3.P

4.O 5.O 6.P

Exercise 5-D

SELECTING SPECIFIC OBJECTIVE-TYPE ITEMS FOR CLASSROOM TESTS

OUTCOME: Identifies the most relevant objective-type items for a given specific learning outcome.

Directions: Indicate the type of objective test item that is most appropriate for measuring each of the specific learning outcomes listed below by circling the appropriate letter to the left of the outcome, using the following key.

KEY A = Short answer B = True-false C = Matching D = Multiple-choice.

A B C D 1. Relates inventors and their inventions.

A B C D 2. Distinguishes between correct and incorrect statements.

A B C D 3. Recalls chemical formulas.

A B C D 4. Identifies the correct date for a historical event.

A B C D 5. Reduces fractions to lowest terms.

A B C D 6. Selects the best reason for an action.

OUTCOME: States specific learning outcomes that can be measured most effectively by each item type.

Directions: For each of the following types of objective test items, state two **specific learning outcomes** that can be most effectively measured by that item type.

Short-answer:

True-false:

Matching:

Multiple-choice:

(Turn up bottom of page for answers)

Answers to SELECTION (5-D)

1.C 2.B 3.A

4.D 5.A 6.D

Exercise 5-E

PREPARING CLASSROOM TESTS AND ASSESSMENTS

OUTCOME: Distinguishes between sound and unsound procedures for construction classroom tests and assessments.

Directions: Indicate whether each of the procedures listed below is **sound (S)** or **unsound (U)** in the construction of classroom tests and assessments by circling the appropriate letter to the left of the statement.

S U 1. Using a table of specification in test preparation.

S U 2. Writing more test items and assessment tasks than needed.

S U 3. Including a large number of items and tasks for each interpretation.

S U 4. Writing items on the day before testing.

S U 5. Including some clues on items to aid slow learners.

S U 6. Putting items and tasks aside for awhile before reviewing them.

OUTCOME: Describes the role of item difficulty in preparing classroom tests.

Directions: Describe what is appropriate item difficulty for tests that are designed for each particular type of interpretation and explain why they differ.

Norm-referenced interpretation:

Criterion-referenced interpretation:

Why they differ:

(Turn up bottom of page for answers)

Answers to SELECTION (5-E)

1.S 2.S 3.S

4.U 5.U 6.S

Exercise 6-A

CHARACTERISTICS OF SHORT-ANSWER, TRUE-FALSE, AND MATCHING ITEMS

OUTCOME: Distinguishes among the characteristics of different item types.

Directions: Indicate which type of objective test item best fits each of the characteristics listed below by circling the appropriate letter, using the following key.

KEY: S = Short answer T = True-false M = Matching

S T M 1. Is classified as a supply-type item.

S T M 2. Most effective where classification is involved.

S T M 3. Is most influenced by guessing.

S T M 4. Is most difficult to score.

S T M 5. Directions are most difficult to write for this type.

S T M 6. Correct answer may be obtained on the basis of misinformation.

OUTCOME: States advantages and limitations of each item type.

Directions: For each of the following types of objective test items, state one advantage and one limitation.

Short-answer item
Advantage:

Limitation:

True-false item
Advantage:

Limitation:

Matching items
Advantage:

Limitation:

(Turn up bottom of page for answers)

Answers to SELECTION (6-A)

1. S 2. M 3. T

4. S 5. M 6. T

Exercise 6-B

EVALUATING AND IMPROVING SHORT-ANSWER ITEMS

OUTCOME: Identifies common faults in short-answer items.

Directions: Indicate the **type of fault**, if any, in each of the following short-answer items by circling the appropriate letter, using the following key.

KEY A = has no faults B = Has more than one correct answer
C = Contains clue to the answer

A B C 1. John Glenn first orbited the earth in _________.

A B C 2. In what year did Borgoyne surrender to Saratoga? ________.

A B C 3. The United Nations Building is located in the city of ______ ________.

A B C 4. An animal that eats only plants is classified as ________.

A B C 5. Test specifications can be indicated by a table of _______.

A B C 6. Abraham Lincoln was born in ________.

OUTCOME: Improves defective short-answer items.

Directions: Rewrite as **well-constructed** short-answer items each of the faulty items in 1-6 above. If an item has no faults, write **no faults** in the space.

1.

2.

3.

4.

5.

6.

(Turn up bottom of page for answers)

Answers to SELECTION (6-B)

1. B 2. A 3. C

4. A 5. C 6. B

Exercise 6-C

EVALUATING AND IMPROVING TRUE-FALSE ITEMS

OUTCOME: Identifies common faults in true-false items.

Directions: Indicate the **type of fault**, if any, in each of the following true-false items by circling the appropriate letter, using the following key.

KEY
A = Has no faults B = Is ambiguous
C = Contains a clue to the answer D= Opinion statement (**not** true or false)

A B C D 1. Camping is fun for the entire family.

A B C D 2. A parasite may provide a useful function.

A B C D 3. The best place to study is in a quiet room.

A B C D 4. A nickel is larger than a dime.

A B C D 5. Abraham Lincoln was born in Kentucky.

A B C D 6. True-false statements should never include the word *always*.

OUTCOME: Improves defective true-false items.

Directions: Rewrite as **well-constructed** true-false items each of the faulty items in 1-6 above. If an item has no faults, write **no faults** in the space.

1.

2.

3.

4.

5.

6.

(Turn up bottom of page for answers)

Answers to SELECTION (6-C)

1. D 2. C 3. D

4. B 5. A 6. C

Exercise 6-D

EVALUATING AND IMPROVING MATCHING ITEMS

OUTCOME: Identifies common faults in a matching exercise.

Directions: Indicate the specific faults in the following matching exercise by circling the appropriate letter below the exercise (Y = yes, N = no).

Directions: Match the items in the two columns.

Column I	*Column II*
____1. Wind vane	A. Used to measure temperature
____2. Tornado	B. Water vapor in the air
____3. Humidity	C. Violent storm
____4. Thermometer	D. Used to measure wind direction

Y N 1. Directions are inadequate.

Y N 2. Columns are inappropriately placed.

Y N 3. Clues are provided in the answers.

Y N 4. List of responses is too short.

Y N 5. Order of responses is improper.

Y N 6. Matching exercise lacks homogeneity.

OUTCOME: Improves a defective matching exercise.

Directions: In the space below, rewrite the matching exercise at the top of the page. You may (1) add material, (2) delete material, or (3) rework it into more than one exercise, if desired. The final product(s) should meet the criteria for a good matching exercise, and, of course, should cover the same type of material.

(Turn up bottom of page for answers)

Answers to SELECTION (6-D)

1. Y 2. Y 3. Y
4. Y 5. Y 6. Y

Exercise 6-E

CONSTRUCTING SHORT-ANSWER, TRUE-FALSE, AND MATCHING ITEMS

OUTCOME: Constructs sample test items that are relevant to stated learning outcomes.

Directions: In the spaces provided, construct (1) two short-answer items (one in question form and one in incomplete statement form), (2) four true-false items, and (3) one four-item matching exercise. State the specific learning outcome for each item or set of items.

Short-answer item (question form)

OUTCOME:

Item:

Short-answer item (incomplete statement form)

OUTCOME:

Item:

True-false items

OUTCOME:

Items:

1.

2.

3.

4.

Matching exercise

OUTCOME:

Directions:

Column I	*Column II*
____1.	
____2.	
____3.	
____4.	

(Why not have the instructor or a fellow student evaluate your exercise?)

Exercise 7-A

CHARACTERISTICS OF MULTIPLE-CHOICE ITEMS

OUTCOME: Identifies the advantages and limitations of multiple-choice items in comparison to other item types.

Directions: The following statements compare multiple-choice (MC) items to other item types with regard to some specific characteristic or use. Indicate whether **test specialists** would **agree (A)** or **disagree (D)** with each statement by circling the appropriate letter.

A D 1. MC items avoid the possible ambiguity of the short-answer item.

A D 2. MC items are easier to construct than true-false items.

A D 3. MC items have less need for homogeneous material than the matching exercise.

A D 4. MC items can be more reliably scored than short-answer items.

A D 5. MC items can measure **all** learning outcomes effectively.

A D 6. MC items have higher reliability per item than true-false items.

OUTCOME: Lists the characteristics of an effective multiple-choice item.

Directions: List the important characteristics of each of the following parts of a multiple-choice item.

Item Stem:

Correct Answer:

Distracters:

(Turn up bottom of page for answers)

Answers to SELECTION (7-A)

1. A 2. D 3. A

4. A 5. D 6. A

Exercise 7-B

EVALUATING STEMS OF MULTIPLE-CHOICE ITEMS

OUTCOME: Distinguishes between effective and ineffective stems for multiple-choice items.

Directions: For each of the following pairs, indicate which element would make the most effective stem for a multiple-choice item by circling its letter (A or B).

1. A Why did the cost of energy rise so rapidly in the 1970s?
   B Which one of the following statements is true about energy?

2. A Achievement tests should
   B Achievement tests are useful for

3. A A whale is a
   B Whales are classified as

4. A Aluminum, which is finding many new uses, is made from
   B Aluminum is made from

5. A The man who first explored Lake Michigan was
   B The Frenchman who first explored Lake Michigan was

6. A Which of the following illustrates what is meant by climate?
   B Which of the following does not illustrate what is meant by climate?

OUTCOME: Describes the faults in ineffective stems for multiple-choice items.

Directions: For each of the ineffective stems in 1-6 above, **briefly describe the type of fault** it contains.

1.

2.

3.

4.

5.

6.

(Turn up bottom of page for answers)

Answers to SELECTION (7-B)

1. A 2. B 3. B

4. B 5. B 6. A

Exercise 7-C

EVALUATING ALTERNATIVES USED IN MULTIPLE-CHOICE ITEMS

OUTCOME: Distinguishes between effective and ineffective use of alternatives in multiple-choice items.

Directions: For each of the following multiple-choice item stems, there are two sets of alternatives. Indicate which set would make the most effective alternatives for the items by circling the letter (A or B). The items are kept simple and the alternatives are placed across the page to save space.

1. A United States astronaut first stepped on the moon in
   A (1) 1967 (2) 1968 (3) 1969 (4) 1970
   B (1) 1959 (2) 1969 (3) 1979 (4) 1989

2. Who was the 33rd President of the United States?
   A (1) Lincoln (2) Wilson (3) Harding (4) Truman
   B (1) Roosevelt (2) Truman (3) Eisenhower (4) Kennedy

3. Which of the following represents what is meant by reforestation?
   A (1) Cutting (2) Replanting (3) Spraying (4) Surveying
   B (1) Recutting (2) Replanting (3) Spraying (4) Resurveying

4. Which of the following best describes observable student performance?
   A (1) Constructs (2) Fears (3) Realizes (4) Thinks
   B (1) Constructs (2) Fears (3) Realizes (4) None of these

OUTCOME: Describes the faults in ineffective sets of alternatives.

Directions: For each of the ineffective sets of alternatives in 1-4 above, briefly describe the type of fault it contains.

1.

2.

3.

4.

(Turn up bottom of page for answers)

Answers to SELECTION (7-C)
1. A 2. B 3. B 4. A

Exercise 7-D

EVALUATING AND IMPROVING MULTIPLE-CHOICE ITEMS

OUTCOME: Identifies common faults in multiple-choice items.

Directions: Indicate the major type of fault, if any, in each of the following multiple-choice items by circling the appropriate letter, using the following key. The alternatives are placed across the page to save space.

KEY
A = No fault
B = Stem is inadequate
C = Contains inappropriate distracters
D = Contains a clue to the answer

A B C D 1. Reliability (a) means consistency (b) is the same as objectivity (c) refers to usability (d) is a synonym for interpretability

A B C D 2. The characteristic that is most desired in test results is (a) consistency (b) reliability (c) stability (d) validity

A B C D 3. If a test is lengthened, its reliability will (a) decrease (b) increase (c) stay the same (d) none of these

A B C D 4. A method of determining reliability that requires correlating scores from two halves of a test is called (a) equivalent forms (b) Kuder-Richardson method (c) split-half method (d) test-retest method

OUTCOME: Improves defective multiple-choice items.

Directions: Rewrite as **well-constructed** multiple-choice items each of the faulty items in 1-4 above. If an item has no faults, write **no faults** in the space.

1.

2.

3.

4.

(Turn up bottom of page for answers)

Answers to SELECTION (7-D)
1. B   2. C   3. C   4. D

Exercise 7-E

CONSTRUCTING MULTIPLE-CHOICE ITEMS

OUTCOME: Constructs sample multiple-choice items that are relevant to stated learning outcomes.

Directions: In some subject area you have studied or plan to teach, construct one multiple-choice item for each of the general instructional objectives listed below.

Knows basic terms

OUTCOME:

Item:

Knows specific facts

OUTCOME:

Item:

Understands principles (or facts)

OUTCOME:

Item:

Applies principles or facts

OUTCOME:

Item:

(Why not let the instructor or a fellow student evaluate your exercise?)

Exercise 8-A

CHARACTERISTICS OF INTERPRETIVE EXERCISES

OUTCOME: Identifies the advantages and limitations of interpretive exercises in comparison to other item types.

Directions: The following statements compare the **interpretive exercise (IE)** to other types with regard to some specific characteristics or use. Indicate whether test specialists would **agree (A)** or **disagree (D)** with the statements by circling the appropriate letter.

A D 1. The IE is more difficult to construct than other item types.

A D 2. The IE can be designed to measure more complex learning outcomes than the single-objective item.

A D 3. The IE provides a more reliable measure of complex learning outcomes than the essay test.

A D 4. The IE is more effective than the essay test for measuring ability to organize ideas.

A D 5. The IE is one of the most effective item types to use with poor readers.

A D 6. The IE measures knowledge of specific facts more effectively than other item types.

OUTCOME: Lists the characteristics of an effective interpretive exercise.

Directions: List the important characteristics of each of the following parts of an interpretive exercise.

Introductory material:

Related test items:

(Turn up bottom of page for answers)

Answers to SELECTION (8-A)

1.A 2.A 3.A

4.D 5.D 6.D

Exercise 8-B

EVALUATING AND IMPROVING THE INTERPRETIVE EXERCISE

OUTCOME: Identifies the common faults in a interpretive exercise.

Directions: Indicate the specific faults in the following interpretive exercise by circling appropriate letter (Y = yes, N = no) by each fault.

INTERPRETIVE EXERCISE

Directions: Read the paragraph and mark your answers.

Some teachers falsely believe that multiple-choice items are limited to the measurement of simple learning outcomes because they depend on the **recognition** of the answer rather than the **recall** of it. However, complex outcomes can be measured, and the selection of the correct answer is not based on the mere recognition of a previously learned answer. It involves the use of higher mental processes to arrive at a solution and then the correct answer is selected from among the alternatives presented. This is the reason multiple-choice items are now called **selection-type** items rather than **recognition-type** items. It makes clear that the answer is selected but that the mental process involved is not limited to recognition.

T F 1. Some teachers think multiple-choice items measure only at the recognition level.
T F 2. All selection-type items are recognition-type items.
T F 3. Multiple-choice items are now called selection-type items.
T F 4. Selection-type items include multiple-choice, true-false, and matching.

Items for evaluating the interpretive exercise:

Y N 1. Directions are adequate.
Y N 2. Some items measure simple reading skill only.
Y N 3. Some items measure extraneous material.
Y N 4. Some of the items can be answered without reading the paragraph.

OUTCOME: Improves defective interpretive exercise.

Directions: Rewrite the directions for the above interpretive exercise and write one true-false item that calls for **interpretation** of the material.

(Turn up bottom of page for answers)

Answers to SELECTION (8-B)

1.N 2.Y 3.Y 4.Y

Exercise 8-C

CONSTRUCTING INTERPRETIVE EXERCISES

OUTCOME: Constructs sample exercise for interpreting a paragraph.

Directions: Construct an interpretive exercise that measure the ability to **interpret a paragraph** of written material. Include complete directions, the paragraph, and at least two multiple-choice items.

(Why not have the instructor or a fellow student evaluate your exercise?)

Exercise 8-D

CONSTRUCTING INTERPRETIVE EXERCISES

OUTCOME: Constructs sample exercise for interpreting pictorial material.

Directions: Construct an interpretive exercise that measures ability to **interpret a picture or cartoon**. Include complete directions, the pictorial material, and **two objective items** of any type.

(Why not have the instructor or a fellow student evaluate your exercise?)

Exercise 8-E

CONSTRUCTING INTERPRETIVE EXERCISES

OUTCOME: Constructs sample exercise for interpreting a table, chart, or graph.

Directions: Construct an interpretive exercise that measures ability to **interpret a table, chart, or graph**. Include complete directions, the pictorial material, and **two objective items** of any type.

(Why not have the instructor or a fellow student evaluate your exercise?)

Exercise 9-A

CHARACTERISTICS OF ESSAY QUESTIONS

OUTCOME: Identifies the advantages and limitations of essay questions in comparison to objective items.

Directions: The following statements compare **essay questions** to **objective items** with regard to some specific characteristic or use. Indicate whether test specialists would **agree (A)** or **disagree (D)** with the statement by circling the appropriate letter.

A D 1. Essay questions are more efficient than objective items for measuring knowledge of facts.

A D 2. Essay questions are more subject to bluffing than objective items.

A D 3. Essay questions are to be favored when measuring ability to organize.

A D 4. Essay questions measure a more limited sampling of content than objective questions in a given amount of testing time.

A D 5. Essay questions provide more reliable scores than objective items.

A D 6. Essay questions can measure complex learning outcomes that are difficult to measure by other means.

OUTCOME: Lists the characteristics of an effective essay question.

Directions: List the important characteristics of each type of essay question.

Restricted-response question:

Extended-response question:

(Turn up bottom of page for answers)

Answers to SELECTION (9-A)

1.D 2.A 3.A

4.A 5.D 6.A

Exercise 9-B

EVALUATING AND IMPROVING ESSAY QUESTIONS

OUTCOME: Describes faults in essay questions and rewrites them as effective items.

Directions: Describe the faults in each of the sample essay questions and rewrite each question so that it meets the criteria for an effective essay item.

1. Why are essay questions better than objective items?

   Faults:

   Rewrite of item:

2. List the rules from your textbook for constructing essay questions.

   Faults:

   Rewrite of item:

3. How do you feel about using essay questions?

   Faults:

   Rewrite of item:

4. Write on one of the following: (1) constructing essay questions, (2) scoring essay questions, (3) using essay questions to improve learning.

   Faults:

   Rewrite of item:

(Why not have the instructor or a fellow student evaluate your exercise?)

Exercise 9-C

CONSTRUCTING RESTRICTED-RESPONSE ESSAY QUESTIONS

OUTCOME: Constructs sample restricted-response essay questions.

Directions: Construct one restricted-response essay question for each of the types of thought questions listed.

1. Comparing two things.

2. Justifying an idea or action.

3. Classifying things or ideas.

4. Applying a fact or principle.

(Why not have the instructor or a fellow student evaluate your exercise?)

Exercise 9-D

CONSTRUCTING EXTENDED-RESPONSE ESSAY QUESTIONS

OUTCOME: Constructs sample extended-response essay questions.

Directions: Construct one extended-response essay question for each of the types of thought questions listed. For each question, describe the scoring procedures to be used and the elements included in the scoring.

1. **Synthesis**: Production of a plan for doing something (e.g., experiment), for constructing something (e.g., graph, table, dress), or for taking some social action (e.g., preventing pollution).

Question:

Scoring procedure:

2. **Evaluation**: Judging the value for something (e.g., a proposal, a book, a poem, a teaching method, a research study) using definite criteria.

Question:

Scoring procedure:

(Why not have the instructor or a fellow student evaluate your exercise?)

Exercise 9-E

SCORING ESSAY QUESTIONS

OUTCOME: Distinguishes between good and bad practices in scoring essay questions.

Directions: Indicate whether each of the following statements describes a **good (G)** practice or a **bad (B)** practice in scoring essay questions by circling the appropriate letter.

G B 1. Use a model answer for scoring restricted-response questions.

G B 2. Evaluate all answers on a student's paper before doing the next paper.

G B 3. Review a student's scores on earlier tests before reading the answers.

G B 4. Score content and writing skills separately.

G B 5. Use the rating method for scoring extended-response questions.

G B 6. Lower the score one point for each misspelled word.

OUTCOME: Prepares a list of points for scoring essay questions.

Directions: Make a list of five do's and five don'ts to serve as a guide for scoring essay tests.

DO:
1.

2.

3.

4.

5.

DON'T:
1.

2.

3.

4.

5.

(Turn up bottom of page for answers)

Answers to SELECTION (9-E)

1.G 2.B 3.B

4.G 5.G 6.B

Exercise 10-A

CHARACTERISTICS OF PERFORMANCE-BASED ASSESSMENT TASKS

OUTCOME: Identifies the advantages and limitations of performance-based assessment tasks.

Directions: The following statements compare **performance-based assessment (PBA) tasks** to **objective items** with regard to some specific characteristic or use. Indicate whether test specialists would **agree (A)** or **disagree (D)** with the statement by circling the appropriate letter.

A D 1. PBA tasks are more likely to model good instruction.

A D 2. PBA tasks provide a better means of assessing the breadth of a student's knowledge.

A D 3. PBA tasks are to be favored when measuring the process that a student uses to solve a problem.

A D 4. PBA tasks measure a more limited sampling of behavior.

A D 5. PBA tasks are more suitable for measuring ability to solve ill-structured problems.

A D 6. PBA tasks provide more reliable scores.

OUTCOME: Lists the characteristics of an effective performance-based assessment task.

Directions: List the important characteristics of each type of performance-based assessment task.

Restricted-response task:

Extended-response task:

(Turn up bottom of page for answers)

Answers to SELECTION (10-A)

1.A 2.D 3.A

4.A 5.A 6.D

Exercise 10-B

CONSTRUCTING RESTRICTED-RESPONSE PERFORMANCE-BASED ASSESSMENT TASKS

OUTCOME: Constructs simple restricted-response performance-based assessment tasks.

Directions: Construct two restricted-response performance-based assessment tasks for a grade and subject matter of your choice. Include a description of the directions to students and the criteria to be used in judging the performances.

1. Directions:

   Task:

   Scoring criteria:

2. Directions:

   Task:

   Scoring criteria:

(Why not let the instructor or a fellow student evaluate your exercise?)

Exercise 10-C

CONSTRUCTING EXTENDED-RESPONSE PERFORMANCE-BASED ASSESSMENT TASKS

OUTCOME: Constructs an extended-response performance-based assessment tasks.

Directions: Construct an extended-response performance-based assessment task involving problem solving for a grade and subject matter of your choice. The task should require students to decide on an approach to solving the problem, identify or gather relevant information, and integrate that information to produce a product. Include a description of the directions to students and the criteria to be used in judging the performances.

Directions:

Task:

Scoring:

(Why not let the instructor or a fellow student evaluate your exercise?)

Exercise 10-D

RATING SCALES

OUTCOME: Distinguishes between desirable and undesirable practices in using rating scales.

Directions: Indicate whether each of the following statements describes a **desirable (D)** practice or an **undesirable (U)** practice in using rating scales with performance-based assessments by circling the appropriate letter.

D U 1. The descriptive graphic rating scale should be favored over the numerical scale.

D U 2. In rating performance, derive the characteristics to be rated from the list of learning objectives.

D U 3. Use a least ten points on each scale to be rated.

D U 4. Use holistic rating procedures to provide students with diagnostic feedback.

D U 5. Separate ratings of secondary characteristics such as neatness from ratings of accomplishment of primary learning objectives.

D U 6. Communicate the criteria to be used in judging performances to students.

OUTCOME: Constructs items for a rating scale.

Directions: Prepare two items for a descriptive graphic rating scale to be used in assessing some type of student performance or some product produced by the student. Do not use sample items from your textbook.

(Turn up bottom of page for answers)

Answers to SELECTION (10-D)

1.D	2.D	3.U
4.U	5.D	6.D

Exercise 10-E

CHECKLISTS

OUTCOME: Distinguishes between desirable and undesirable practices in using checklists.

Directions: Indicate whether each of the following statements describes a **desirable (D)** practice or an **undesirable (U)** practice in using checklists by circling the appropriate letter.

D U 1. Use a checklist wherever frequency of occurrence is an important element in the assessment.

D U 2. In assessing performance, include in a checklist both desired actions and common errors.

D U 3. Use a checklist for assessing some products.

D U 4. Use a checklist to determine if steps in performance were completed in proper order.

D U 5. Use a checklist for assessing process but not for assessing student products.

D U 6. Avoid use of checklists in assessing process.

OUTCOME: Constructs a performance checklist.

Directions: Prepare a brief checklist for some simple performance-based assessment task. Include directions describing how to respond.

(Turn up bottom of page for answers)

Answers to SELECTION (10-E)

1.U   2.D   3.D

4.D   5.U   6.U

Exercise 11-A

ANECDOTAL RECORDS

OUTCOME: Distinguishes between desirable and undesirable practices in using anecdotal records.

Directions: Indicate whether each of the following statements describes a **desirable (D)** practice or an **undesirable (U)** practice in using anecdotal records by circling the appropriate letter.

D U 1. Confine observations to areas that can be verified by objective testing.

D U 2. Keep factual descriptions of incidents and interpretations of them separate.

D U 3. Limit each anecdote to a single incident.

D U 4. Limit each anecdote to the behavior of only one student.

D U 5. Wait until after school hours to record the observed incidents.

D U 6. Record both positive and negative incidents.

OUTCOME: Writes an anecdotal record on an incident.

Directions: Briefly observe some aspect of student performance (e.g., speaking, playing a game) and write an anecdotal record of the incident.

(Turn up bottom of page for answers)

Answers to SELECTION (11-A)

1. U 2. D 3. D

4. U 5. U 6. D

Exercise 11-B

USE OF PEER APPRAISAL AND SELF-REPORT TECHNIQUES

OUTCOME: Selects the most appropriate technique for a particular use.

Directions: Indicate which technique is **most appropriate** for each of the uses listed below by circling the appropriate letter.

KEY

G = "Guess who" technique S = Sociometric technique
A = Attitude scale I = Interest inventory

G S A I 1. To analyze the social structure of a group.

G S A I 2. To aid in selecting reading material for a poor reader.

G S A I 3. To determine the reputation a student holds among his or her classmates.

G S A I 4. To aid in assessing the effectiveness of instructional activities.

G S A I 5. To determine how well a particular student is accepted by his or her classmates.

G S A I 6. To aid students in career planning.

OUTCOME: States the advantages and disadvantages of peer appraisal and self-report techniques.

Directions: Briefly state **one advantage** and **one disadvantage** of each of the following techniques.

Peer appraisal:

Self report:

(Turn up bottom of page for answers)

Answers to SELECTION (11-B)

1. S 2. I 3. G

4. A 5. S 6. I

Exercise 11-C

GUESS WHO TECHNIQUE

OUTCOME: Distinguishes between desirable and undesirable practices in using the guess who technique.

Directions: Indicate whether each of the following statements describes a **desirable (D)** practice or an **undesirable (U)** practice in using the guess who techniques by circling the appropriate letter.

D U 1. Use only clearly favorable behavior descriptions.

D U 2. Have students write as many names as they wish for each behavior description.

D U 3. Permit students to name a person for more than one behavior description.

D U 4. Have students respond by using first name and initial of last name.

D U 5. Use the guess who technique for assessing personal and social development only.

D U 6. Score the responses by counting the number of nominations a student receives on each behavior description.

OUTCOME: Constructs items for a guess who form.

Directions: List six statements that could be used in a guess who form for evaluating students "study and work habits."

(Turn up bottom of page for answers)

Answers to SELECTION (11-C)

1. U   2. D   3. D

4. D   5. U   6. D

Exercise 11-D

SOCIOMETRIC TECHNIQUE

OUTCOME: Distinguishes between desirable and undesirable practices in using the sociometric technique.

Directions: Indicate whether each of the following statements describes a **desirable (D)** practice or and **undesirable (U)** practice in using the sociometric technique by circling the appropriate letter.

D U 1. Sociometric choices should be real choices that are actually used to arrange groups.

D U 2. The situations used in sociometric choosing should be ones in which all students are equally free to participate.

D U 3. Students should be told to state their first choice **only**, in order to simplify the tabulation of results.

D U 4. The plotted sociogram should show the social position of each student and the social pattern of the group.

D U 5. Each student should be shown, in an individual conference, his or her place on the sociogram.

D U 6. Sociometric choices should be used to assess the influence of school practices on students' social relations.

OUTCOME: Constructs items for a sociometric form.

Directions: List three choice situations to be used on a sociometric form. The situations should be suitable for the grade level at which it is to be used. Indicate the grade level. Do not use any sample items from your textbook.

(Turn up bottom of page for answers)

Answers to SELECTION (11-D)

1. D	2. D	3. U
4. D	5. U	6. D

Exercise 11-E

ATTITUDE MEASUREMENT

OUTCOME: Distinguishes between desirable and undesirable practices in using a Likert-type attitude scale.

Directions: Indicate whether each of the following statements describes a **desirable (D)** practice or an **undesirable (U)** practice in using a Likert-type attitude scale by circling the appropriate letter.

D U 1. Use only clearly favorable and unfavorable attitude statements.

D U 2. Have students write statements for use in the attitude scale.

D U 3. Arrange the statements in the scale in order, from least favorable to most favorable.

D U 4. Have students respond by indicating how strongly they agree or disagree.

D U 5. Use a group of judges to obtain scoring weights.

D U 6. Have students put their names on the attitude scale.

OUTCOME: Constructs a Likert-type attitude scale.

Directions: List six statements that could be used to measure student attitudes toward **testing** according to a Likert-type scale. Include a place to respond and the scoring weights for each item.

(Turn up bottom of page for answers)

Answers to SELECTION (11-E)

1. D   2. D   3. U

4. D   5. U   6. U

Exercise 12-A

REVIEWING AND ARRANGING ITEMS AND TASKS IN CLASSROOM TESTS AND ASSESSMENTS

OUTCOME: Distinguishes between good and bad practices in reviewing and arranging test items and assessment tasks.

Directions: Indicate whether each of the following statements describes a **good (G)** practice or a **bad (B)** practice in reviewing items and tasks and arranging them in classroom tests and assessments by circling the appropriate letter.

G B 1. Have another teacher review the items and tasks for defects.

G B 2. Recheck relevance to the specific learning outcome when reviewing an item or task.

G B 3. During item and task review, remove any racial or sexual stereotyping.

G B 4. In arranging items in the test, place essay questions or other performance tasks first.

G B 5. Intersperse true-false items among multiple-choice items.

G B 6. Put easy items last to maintain student motivation.

OUTCOME: Prepares a list of points for reviewing and arranging test items and assessment tasks.

Directions: Make a list of **four do's** and **four don'ts** to serve as a guide for reviewing and arranging test items and assessment tasks.

DO

1.

2.

3.

4.

DON'T

1.

2.

3.

4.

(Turn up bottom of page for answers)

Answers to SELECTION (12-A)

1.G 2.G 3.G

4.B 5.B 6.B

Exercise 12-B

PREPARING TEST DIRECTIONS

OUTCOME: Prepares sample directions for a classroom test.

Directions: Prepare a complete set of directions for a test in some subject. Assume that the general directions for the test as a whole and the specific directions for each item type are all for the same test.

General directions:

Short-answer items:

True-false items:

Multiple-choice items:

Matching items:

Why not have the instructor or a fellow student evaluate your exercise?

Exercise 12-C

ADMINISTERING AND SCORING CLASSROOM TESTS AND ASSESSMENTS

OUTCOME: Distinguishes between good and bad practice in administering and scoring classroom tests and assessments.

Directions: Indicate whether each of the following statements describes a **good (G)** practice or a **bad (B)** practice in administering and scoring classroom tests and assessments by circling the appropriate letter.

G B 1. Students are told whether there is a correction for guessing.

G B 2. Students are told how much time to spend on each part of the test and each extended response task.

G B 3. Students are told to skip items that seem too difficult and come back to them later.

G B 4. The teacher explains the meaning of an ambiguous question to the student who asked about it.

G B 5. An objective test is scored by counting important items 1 point and very important items 2 points.

G B 6. Students answer every item, so their scores are corrected for guessing.

OUTCOME: Describes and illustrates use of the correction-for-guessing formula.

Directions: Describe when the correction-for-guessing formula should and should not be used for classroom tests and compute the corrected scores for the given data.

Use the correction formula when:

Do not use the correction formula when:

Compute the corrected scores on an eight-item, true-false test for the student responses shown below: (R = Right, W = Wrong, O = Omit)

	1	2	3	4	5	6	7	8	Corrected Score
Bob	R	R	R	R	R	R	O	O	
Sara	R	R	R	R	R	W	R	W	
Terry	R	R	R	W	O	R	W	W	

(Turn up bottom of page for answers)

Answers to SELECTION (12-C)

1.G 2.G 3.G

4.B 5.B 6.B

Bob, 6; Sara, 4; Terry, 1

Exercise 12-D

ITEM ANALYSIS PROCEDURES FOR NORM-REFERENCED CLASSROOM TESTS

OUTCOME: Computes and interprets item analyses.

Directions: Review the following item analysis data and answer the questions below the data. For each item there are 10 students in the upper group and 10 students in the lower group. The data indicate the number of students choosing each alternative. The correct answer is marked (*).

Item number		Alternatives A	B	C	D
1.	Upper	*10	0	0	0
	Lower	*2	2	3	3
2.	Upper	0	1	*8	1
	Lower	5	1	*3	1
3.	Upper	2	*5	2	1
	Lower	3	*4	2	1
4.	Upper	*6	2	1	1
	Lower	*2	3	2	3
5.	Upper	3	0	0	*7
	Lower	0	4	5	*1

1. What is the **difficulty** of item 2? ________

2. What is the **discriminating** power of item 2? _________

3. Which one of the above items is **most** discriminating? ________

4. Which one of the above items is **least** discriminating? ________

5. Which one of the above items has a **negatively** discriminating distracter? ________

6. Which one of the above items most closely meets **all** the criteria for an ideal item? ________

List the cautions in interpreting item analysis results like the above.

(Turn up bottom of page for answers)

Answers to SELECTION (12-D)

1. 55% 2. .50 3. 1

4. 3 5. 5 6. 1

Exercise 12-E

APPLICATION OF ITEM ANALYSIS PRINCIPLES TO PERFORMANCE-BASED ASSESSMENT TASKS

OUTCOME: Applies and interprets item analysis principles with performance-based assessment tasks.

Directions: A set of eight performance-based assessment tasks were administered to a group of 30 students. Each task was scored on a five-point scale. The total score for the assessment was the sum of the eight task scores. Total scores for the assessment and scores on task 8 of the assessment are listed below. The scores are listed in order of the total score. Use these data to analyze the discriminating power of task 9 by comparing the performance of the upper and lower groups of ten students.

Student	1	2	3	4	5	6	7	8	9	10
Total Score	36	35	35	34	33	33	33	32	32	32
Score on Task 8	5	5	4	5	4	4	4	3	5	3

Student	11	12	13	14	15	16	17	18	19	20
Total Score	31	30	30	30	29	29	29	29	28	28
Score on Task 8	4	4	3	3	4	3	3	3	3	2

Student	21	22	23	24	25	26	27	28	29	30
Total Score	27	27	26	25	23	23	21	20	18	16
Score on Task 8	4	3	3	3	3	2	3	2	1	1

Construct an analysis table for Item 8.

Briefly interpret the results of your analysis.

(Turn up bottom of page for answers)

Answers to SELECTION (12-E)

Score	1	2	3	4	5	Average
Upper	0	0	2	4	4	4.2
Lower	2	2	5	1	0	2.5

Exercise 13-A

TYPES OF MARKING AND REPORTING SYSTEMS

OUTCOME: Distinguishes among the characteristics of different types of marking and reporting systems.

Directions: Indicate which type of marking and reporting system best fits each statement listed below by circling the appropriate letter, using the following key.

KEY: A = Traditional letter grade (A,B,C,D,F) B = Two-letter grade (pass, fail)
C = Checklist of objectives D = Parent-teacher conference

A B C D 1. Provides for two-way reporting.

A B C D 2. Provides most useful learning guide to student.

A B C D 3. Provides least information concerning learning.

A B C D 4. Most preferred by college admissions officers.

A B C D 5. May be too complex to be understood by parents.

A B C D 6. Most widely used method of reporting in high school.

OUTCOME: Lists the advantages and disadvantages of the traditional (A,B,C,D,F) marking system.

Directions: List the advantages and disadvantages of using the traditional (A,B,C,D,F) marking system as the **sole** method of reporting student progress.

Advantages:

Disadvantages:

(Turn up bottom of page for answers)

Answers to SELECTION (13-A)

1.D 2.C 3.B

4.A 5.C 6.A

Exercise 13-B

ASSIGNING RELATIVE LETTER GRADES

OUTCOME: Distinguishes between desirable and undesirable practices in assigning relative letter grades.

Directions: Indicate whether each of the following statements describes a **desirable (D)** practice or an **undesirable (U)** practice in assigning relative letter grades by circling the appropriate letter.

D U 1. The grades should reflect the learning outcomes specified for the course.

D U 2. To give test scores equal weight in a composite score, the scores should be simply added together.

D U 3. If it is decided to assign different weights to some scores, the weighting should be based on the maximum possible score on the test.

D U 4. Grades should be lowered for tardiness or misbehavior.

D U 5. Grading typically should be based on the normal curve.

D U 6. Pass-fail decisions should be based on an absolute standard of achievement.

OUTCOME: Assigns weights in obtaining composite scores for grading purposes.

Directions: Following is a list of types of information a teacher would like to include in assigning a final grade to each student. If the teacher wants to count each type of information one-fourth of the final grade, what weight should be given to each type of information?

Type of Information	Range of Scores	Weight to Be Used
Midsemester examination	30 to 50	
Term project	5 to 10	
Performance assessments	15 to 25	
Final examination	20 to 100	

(Turn up bottom of page for answers)

Answers to SELECTION (13-B)

1.D 2.U 3.U

4.U 5.U 6.D

Exercise 13-C

ASSIGNING ABSOLUTE GRADES

OUTCOME: Distinguishes between desirable and undesirable practices in assigning absolute grades.

Directions: Indicate whether each of the following statements describes a **desirable (D)** practice or an **undesirable (U)** practice in assigning absolute letter grades by circling the appropriate letter.

D U 1. Absolute grades should be used with mastery learning.

D U 2. Clearly defined domains of learning tasks should provide the basis for grading.

D U 3. If all students pass a test, a harder test should be given before grades are assigned.

D U 4. The distribution of grades to be assigned should be predetermined and explained.

D U 5. Grades should be based on the amount of improvement shown.

D U 6. When using absolute grading the standard for passing should be predetermined.

OUTCOME: Lists guidelines for effective grading.

Directions: List five important guidelines for effective grading.

1.

2.

3.

4.

5.

(Turn up bottom of page for answers)

Answers to SELECTION (13-C)

1.D 2.D 3.U

4.U 5.U 6.D

Exercise 13-D

PARENT-TEACHER CONFERENCE

OUTCOME: Distinguishes between desirable and undesirable practices in conducting a parent-teacher conference.

Directions: Indicate whether each of the following statements describes a **desirable (D)** practice or an **undesirable (U)** practice in conducting parent-teacher conferences by circling the appropriate letter.

D U 1. Before the conference, assemble a portfolio of specific information about and examples of the student's learning progress.

D U 2. Present examples of the student's work to parents.

D U 3. Begin the conference by describing the student's learning difficulties.

D U 4. Make clear to parents that, as a teacher, you know what is best for the student's learning and development.

D U 5. In the concluding phase, review your conference notes with the parents.

D U 6. End the conference with a positive comment about the student.

OUTCOME: Lists questions that might be asked of parents during the conference.

Directions: Write a list of questions that you could ask parents during a conference that might help you better understand students' problems of learning and development.

(Turn up bottom of page for answers)

Answers to SELECTION (13-D)

1.D 2.D 3.U

4.U 5.D 6.D

Exercise 13-E

REPORTING RESULTS OF PUBLISHED TESTS TO PARENTS

OUTCOME: Distinguishes between desirable and undesirable practices in reporting results of published tests to parents.

Directions: Indicate whether each of the following statements describes a **desirable (D)** practice or an **undesirable (U)** practice in reporting results of published tests to parents by circling the appropriate letter.

D U 1. Describe what the test measures in brief, understandable terms.

D U 2. Make clear the distinction between percentile rank and percentage-correct scores.

D U 3. Use grade equivalent scores to indicate the grade at which the student can perform.

D U 4. Use the same type of score to report all test results, whenever possible.

D U 5. Describe a difference between two test scores as a "real difference" only after the error of measurement is considered.

D U 6. Explain how the test results will be used only if the parent asks.

OUTCOME: Identifies and corrects common errors in reporting standardized test results.

Directions: Indicate what is wrong with each of the following statements and rewrite each one so that it provides an accurate report.

1. "Derek's percentile rank of 70 in spelling means he can spell 70 percent of the words in the test."

2. "Marie's stanine score of 6 in reading indicates she is performing below average in reading."

3. "Erik's grade-equivalent scores of 5.4 in reading and 6.2 in math indicate that his performance in math is superior to his performance in reading."

(Turn up bottom of page for answers)

Answers to SELECTION (13-E)

1.D 2.D 3.U

4.D 5.D 6.U

Exercise 14-A

STANDARDIZED ACHIEVEMENT TESTS VERSUS INFORMAL CLASSROOM TESTS

OUTCOME: Identifies the comparative advantages of standardized and informal classroom tests for measuring student achievement.

Directions: Indicate whether each of the following statements best describes a **standardized** achievement test **(S)** or an informal **classroom** test **(C)** by circling the appropriate letter.

S C 1. Likely to be more relevant to a teacher's instructional objectives.

S C 2. Likely to provide more reliable test scores.

S C 3. Technical quality of test items is consistently high.

S C 4. Most useful in formative assessment.

S C 5. Typically provides the larger spread of scores.

S C 6. Best for use in rapidly changing content areas.

OUTCOME: States a major advantage and limitation of standardized achievement tests.

Directions: Briefly state one major advantage and one major limitation of standardized achievement tests.

Advantage:

Limitation:

(Turn up bottom of page for answers)

Answers to SELECTION (14-A)

1.C 2.S 3.S

4.C 5.S 6.C

Exercise 14-B

USE OF PUBLISHED ACHIEVEMENT TEST BATTERIES

OUTCOME: Selects the most appropriate type of achievement test battery for a particular use.

Directions: Indicate which type of test is most useful for each of the following testing purposes by circling the appropriate letter using the following key.

KEY S = Survey achievement test battery
D = Diagnostic achievement test battery

S D 1. To compare schools on basic skill development.

S D 2. To describe the specific skills a student has yet to learn in reading.

S D 3. To measure achievement in science and social studies.

S D 4. To detect specific weaknesses in adding fractions.

S D 5. To determine how a fifth-grade class compares to other fifth grade classes in reading.

S D 6. To determine mastery of particular language skills.

OUTCOME: States a major advantage and limitation of achievement test batteries.

Directions: Briefly state one major advantage and one major limitation of achievement test batteries of the **survey** type.

Advantage:

Limitation:

(Turn up bottom of page for answers)

Answers to SELECTION (14-B)

1.S 2.D 3.S

4.D 5.S 6.D

Exercise 14-C

COMPARISON OF READING READINESS TESTS AND READING SURVEY TESTS

OUTCOME: Identifies the functions measured by different types of reading tests.

Directions: Indicate whether each of the functions listed below is measured by reading **readiness** tests **(R)**, by reading **survey** tests **(S)**, by **both (B)**, or by **neither (N)**, by circling the appropriate letter.

R S B N 1. Auditory discrimination.

R S B N 2. Comprehension of the meaning of words.

R S B N 3. Ability to draw inferences.

R S B N 4. Ability to read maps.

R S B N 5. Rate of reading.

R S B N 6. Attitude toward reading.

OUTCOME: Compares reading tests.

Directions: Compare two reading survey tests (or readiness tests) for a particular grade level (1) briefly describe how the two tests differ, and (2) indicate which one you would prefer to use for a particular purpose and why.

(Turn up bottom of page for answers)

Answers to SELECTION (14-C)

1.R 2.B 3.S

4.N 5.S 6.N

Exercise 14-D

COMPARISON OF STANDARDIZED AND CUSTOMIZED ACHIEVEMENT TESTS

OUTCOME: Distinguishes among the characteristics of different types of achievement tests.

Directions: Indicate which type of test best fits each feature listed below by circling the appropriate letter, using the following key.

KEY: S = Standardized Achievement Tests
C = Customized Achievement Tests

S C 1. Are most useful to the classroom teacher.

S C 2. Have the greatest need for adequate norms.

S C 3. Most adaptable to changing conditions.

S C 4. Most likely to have some content the students have not studied.

S C 5. Best for making criterion-referenced interpretations.

S C 6. Likely to provide the most valid measure of local instructional objectives.

OUTCOME: Describes the procedure for producing customized achievement tests.

Directions: List and briefly describe the procedural steps to follow in producing locally prepared customized achievement tests.

(Turn up bottom of page for answers)

Answers to SELECTION (14-D)

1.C 2.S 3.C

4.S 5.C 6.C

Exercise 14-E

SELECTING PUBLISHED ACHIEVEMENT TESTS

OUTCOME: Selects the type of test that is most appropriate for a particular purpose.

Directions: For each of the following statements, indicate which type of test would be used by circling the appropriate letter using the following key.

KEY: A = Achievement Test Battery B = Separate Test of Content
C = Customized Achievement Test D = Individual Achievement Test

A B C D 1. To test student's mastery of classroom objectives.

A B C D 2. To test a student who has a learning disability.

A B C D 3. To compare a student's performance in reading and mathematics.

A B C D 4. To test students at the end of each unit.

A B C D 5. To determine how a classroom final examination in science compares to a standardized test.

A B C D 6. To measure student progress from one grade level to the next.

OUTCOME: Compares the usefulness of customized tests and standardized tests.

Directions: State one advantage and one disadvantage of using a customized test instead of a standardized test to measure student achievement.

Advantage:

Disadvantage:

(Turn up bottom of page for answers)

Answers to SELECTION (14-E)

1.C 2.D 3.A

4.C 5.B 6.A

Exercise 15-A

COMPARISON OF APTITUDE AND ACHIEVEMENT TESTS

OUTCOME: Identifies the similarities and differences in the characteristics of aptitude and achievement tests.

Directions: Indicate whether each of the following statements is characteristic of an **aptitude (P)** test, an **achievement (C)** test, or **both (B)** types of tests by circling the appropriate letter.

P C B 1. Measures learned ability.

P C B 2. Useful in predicting future achievement.

P C B 3. Content-related evidence of validity is emphasized.

P C B 4. Criterion-related evidence of validity is emphasized.

P C B 5. Can be used in grades from kindergarten through grade 12.

P C B 6. Emphasizes reasoning abilities.

OUTCOME: Lists the major differences between aptitude and achievement tests.

Directions: List the **major differences** between aptitude tests and achievement tests.

(Turn up bottom of page for answers)

Answers to SELECTION (15-A)

1.B   2.B   3.C

4.P   5.B   6.P

Exercise 15-B

GROUP TESTS OF LEARNING ABILITY

OUTCOME: Identifies the types of scores provided by selected group tests.

Directions: Indicate the types of scores provided by each of the group tests listed below by circling the appropriate letter using the following key.

KEY: A = single score, B = verbal and quantitative scores only, C = verbal, nonverbal and total scores only, D = verbal, quantitative, and nonverbal scores, E = more than three scores.

A B C D E 1. Cognitive Abilities Test.

A B C D E 2. Differential Aptitude Tests.

A B C D E 3. Matrix Analogies Test.

A B C D E 4. Otis-Lennon School Ability Test.

A B C D E 5. School and College Ability Test.

OUTCOME: States advantages and disadvantages of using different types of learning ability tests.

Directions: Briefly state one advantage and one disadvantage of each type of group test of learning ability.

Single score:

Separate-scores (verbal, nonverbal, quantitative):

(Turn up bottom of page for answers)

Answers to SELECTION (15-B)

1.D 2.E 3.A

4.C 5.B

Exercise 15-C

INDIVIDUAL TESTS

OUTCOME: Identifies the similarities and differences in the characteristics of individual tests.

Directions: Indicate whether each of the following statements is characteristic of the Stanford-Binet Intelligence Scale **(S)**, the Wechsler Intelligence Scales-Revised **(W)**, or **both (B)**.

S W B 1. Uses a variety of item types.

S W B 2. Items are arranged by subtest.

S W B 3. Includes a vocabulary test.

S W B 4. Provides separate verbal and performance IQs.

S W B 5. Scores are reported in Standard Age Scores.

S W B 6. Provides total scores and scores on subtests.

OUTCOME: List conditions that might lower scores on tests of learning abilities.

Directions: List five conditions that might lower a student's score on a test of learning ability.

1.

2.

3.

4.

5.

(Turn up bottom of page for answers)

Answers to SELECTION (15-C)

1.B 2.B 3.B

4.W 5.S 6.B

Exercise 15-D

DIFFERENTIAL APTITUDE TESTING

OUTCOME: Identifies the characteristics of the Differential Aptitude Tests (DAT).

Directions: Indicate whether each of the following statements is characteristic of the Differential Aptitude Tests (DAT) by circling **yes** (if it is) and **no** (if it is not).

Yes No 1. The DAT would be classified as a test battery.

Yes No 2. The intercorrelations between subtests on the DAT are high (average about .90).

Yes No 3. Some of the DAT subtests measure abilities like those measured by group scholastic aptitude tests.

Yes No 4. The DAT profile indicates scores in terms of percentile rank.

Yes No 5. The DAT can be administered "adaptively."

Yes No 6. The eight tests on the DAT are speed tests.

OUTCOME: States a major advantage and limitation of the Differential Aptitude Tests.

Directions: Briefly state one major advantage and one major limitation of using the Differential Aptitude Tests instead of a series of separate tests from different publishers.

Advantage of DAT:

Limitation of DAT:

(Turn up bottom of page for answers)

Answers to SELECTION (15-D)

1.Y 2.N 3.Y

4.Y 5.Y 6.N

Exercise 15-E

SELECTING APPROPRIATE TESTS

OUTCOME: Selects the type of test that is most appropriate for a particular use.

Directions: For each of the following purposes indicate which type of test should be used by circling the appropriate letter, using the following key.

KEY: G = Group test of learning ability I = Individual test of learning ability
D = Differential aptitude tests

G I D 1. To test a preschool child.

G I D 2. To test a fourth-grade student who is unable to speak.

G I D 3. To test a sixth-grade student who has a severe learning disability.

G I D 4. To assist a tenth-grade student with career planning.

G I D 5. To aid forming learning groups within the classroom.

G I D 6. To aid in planning an individual program for students with severe learning disabilities.

OUTCOME: Compares the usefulness of culture-fair test and conventional test of learning ability.

Directions: State one advantage and one disadvantage of using a culture-fair test instead of a conventional learning ability test for testing students from disadvantaged homes.

Advantage:

Disadvantage:

(Turn up bottom of page for answers)

Answers to SELECTION (15-E)

1.I 2.G 3.I

4.D 5.G 6.I

Exercise 16-A

SOURCES OF INFORMATION ON PUBLISHED TESTS

OUTCOME: Identifies the most useful source of information for a given situation.

Directions: Below is a list of four sources of information concerning published tests. For each of the statements following the list, indicate the source of information that should be consulted **first** by circling the appropriate letter.

KEY

A = Mental Measurements Yearbooks B = Professional journals
C = Test manual D = Test publisher's catalog

A B C D 1. To locate the most recent tests in an area.

A B C D 2. To obtain critical reviews of a published test.

A B C D 3. To find out how a particular published test was constructed.

A B C D 4. To locate the most recent research studies using a particular test.

A B C D 5. To determine if any tests of study skills have been published.

A B C D 6. To determine the type of norms used in a published test.

OUTCOME: Summarizes the purpose and content of the Standards for Educational and Psychological Testing.

Directions: Briefly describe the purpose and content of the Standards for Educational and Psychological Testing.

Purpose:

Content:

(Turn up bottom of page for answers)

Answers to SELECTION (16-A)

1.D 2.A 3.C

4.B 5.A 6.C

Exercise 16-B

EVALUATING AN ACHIEVEMENT TEST

OUTCOME: Evaluates a test using a test evaluation form.

Directions: Select an achievement test at the grade level of your choice and obtain a copy of the test, the manual, and other accessory material; your instructor can help you with this. Study the test materials, consult the reviews in the latest Mental Measurements Yearbook (MMY) and Test Critiques, and write your evaluation using the following test evaluation form. Be brief and include only the most essential information.

TEST EVALUATION FORM

Test title ______________________________ Author(s)______________________________

Publisher ______________________________ Copyright date(s) ______________________________

Purpose of test ____

For grades (ages)______________________________ Forms ______________________________

Scores available ______________________________ Method of scoring ______________________________

Administration time______________________________ Time(s) of parts ______________________________

Validity (cite manual pages) ____________. Summarize evidence below.

Content considerations:

Test-criterion relationships:

Construct considerations:

Evidence regarding consequences of use:

TEST EVALUATION FORM CONTINUED

Reliability (cite manual page numbers) ______. Summarize data below.

Age or Grade	Type of reliability	Range of reliabilities (Total test)	Number Tested	Range of Reliabilities (Part scores)

Standard errors of measurement ___________ ___________

Norms (cite manual page numbers) __________. Summarize data below.

Type (e.g., percentile rank):

Groups (size, age, or grade):

Separate norms (e.g., type of district):

Criterion-referenced interpretation

Describe (if available):

Practical features

Ease of administration:

Ease of scoring:

Ease of interpretation:

Adequacy of manual and materials:

Comments of reviewers (See MMY or Test Critiques)

Summary Evaluation

Advantages:

Limitations:

Exercise 16-C

EVALUATING AN ABILITY TEST

OUTCOME: Evaluates a test using test evaluation form.

Directions: Select an aptitude test at the grade level of your choice and obtain a copy of the test, the manual, and other accessory material; your instructor can help you with this. Study the test materials, consult the reviews in the latest Mental Measurements Yearbook (MMY) and Test Critiques, and write your evaluation using the following test evaluation form. Be brief and include only the most essential information.

TEST EVALUATION FORM

Test title ______________________ Author(s)______________________________

Publisher ______________________ Copyright date(s) ________________________

Purpose of test ____

For grades (ages)_________________ Forms ________________________________

Scores available ________________ Method of scoring _______________________

Administration time_______________ Time(s) of parts _______________________

Validity (cite manual pages) _________. Summarize evidence below.

Content considerations:

Test-criterion relationships:

Construct considerations:

Evidence regarding consequences of use:

TEST EVALUATION FORM CONTINUED

Reliability (cite manual page numbers) _______. Summarize data below.

Age or Grade	Type of reliability	Range of reliabilities (Total test)	Number Tested	Range of reliabilities (Part scores)

Standard errors of measurement ____________ ____________

Norms (cite manual page numbers) ___________. Summarize data below.

Type (e.g., percentile rank):

Groups (size, age, or grade):

Separate norms (e.g., type of district):

Criterion-referenced interpretation

Describe (if available):

Practical features

Ease of administration:

Ease of scoring:

Ease of interpretation:

Adequacy of manual and materials:

Comments of reviewers (See MMY or Test Critiques)

Summary Evaluation

Advantages:

Limitations:

Exercise 16-D

ADMINISTERING PUBLISHED TESTS

OUTCOME: Distinguishes between good and bad practices in administering published tests.

Directions: Indicate whether each of the following statements describes a **good (G)** practice or a **bad (B)** practice in administering a published test by circling the appropriate letter.

G B 1. Read the directions word for word.

G B 2. Give students extra time if there was an interruption during testing.

G B 3. If the directions didn't seem clear, tell the students again how to record their answers.

G B 4. Tell students what to do about guessing if the directions failed to include it.

G B 5. If asked about a particular item, tell the student: "I'm sorry but I cannot help you. Do the best you can."

G B 6. Make a record of any unusual student behavior during testing.

OUTCOME: Describes a procedure for improving students' test-taking skills.

Directions: Briefly describe an ethical procedure a classroom teacher might follow for improving students' test-taking skills.

(Turn up bottom of page for answers)

Answers to SELECTION (16-D)

1.G   2.B   3.G

4.B   5.G   6.G

Exercise 16-E

USES OF PUBLISHED TESTS

OUTCOME: Distinguishes between correct and incorrect statements concerning uses of published tests.

Directions: Indicate whether test specialists would **agree (A)** or **disagree (D)** with each of the following statements concerning **test use** by circling the appropriate letter.

A D 1. Published achievement tests are most useful in the areas of basic skills.

A D 2. Published achievement tests need **not** match the instructional objectives of the school.

A D 3. To detect underachievement, relatively large differences are needed between the scores of learning ability tests and achievement tests.

A D 4. Norm-referenced achievement tests are especially useful for individualizing instruction.

A D 5. Course grades are more valid when based on scores from published achievement tests.

A D 6. No important educational decision should be based on the scores of published tests alone.

OUTCOME: Lists misuses of published tests.

Directions: List as many ways as you can think of that published test results might be misused. Use brief concise statements.

(Turn up bottom of page for answers)

Answers to SELECTION (16-E)

1.A 2.D 3.A

4.D 5.D 6.A

Exercise 17-A

USES OF CRITERION-REFERENCED AND NORM-REFERENCED INTERPRETATIONS

OUTCOME: Relates test interpretation to the type of information needed.

Directions: For each of the following questions, indicate whether a **criterion-referenced (C)** or a **norm-referenced (N)** interpretation would be more useful by circling the appropriate letter.

C N 1. How does a student's test performance compare to that of other students in the same grade?

C N 2. What type of remedial work would be most helpful for a slow-learning student?

C N 3. How does a student's test performance in reading and mathematics compare?

C N 4. Which students' test performances exceed those of 90 percent of their classmates?

C N 5. Which students have achievement mastery of computational skills?

C N 6. How does student test performance in our school compare with that of other schools?

OUTCOME: Describes the cautions needed when making criterion-referenced interpretations of tests designed for norm-referenced use.

Directions: List and briefly describe several factors to consider when using criterion-referenced interpretations with norm-referenced survey tests.

(Turn up bottom of page for answers)

Answers to SELECTION (17-A)

1.N 2.C 3.N

4.N 5.C 6.N

Exercise 17-B

NATURE OF DERIVED SCORES

OUTCOME: Distinguishes among the characteristics of different types of derived scores.

Directions: Indicate which type of derived score is described by each statement listed below by circling the appropriate letter. Use the following key.

KEY: G = grade equivalent scores P = percentile rank S = standard scores

G P S 1. Provides units that are based on the average score earned in different groups.

G P S 2. Provides units that are **systematically unequal**.

G P S 3. Provides units that are most nearly equal.

G P S 4. Provides units that are most meaningful when interpreted with reference to normal curves.

G P S 5. Provides units that are most meaningful at the elementary school level.

G P S 6. Provides units that are easily interpreted and typically compare students with their own age group.

OUTCOME: States the advantages and limitations of derived scores.

Directions: Briefly state one advantage and one limitation of percentile scores and standard scores.

Percentile Ranks
Advantage:

Disadvantage:

Standard Scores
Advantage:

Disadvantage:

(Turn up bottom of page for answers)

Answers to SELECTION (17-B)

1.G 2.P 3.S

4.S 5.G 6.P

Exercise 17-C

GRADE EQUIVALENT SCORES

OUTCOME: Distinguishes between appropriate and inappropriate interpretations of grade equivalent scores.

Directions: Indicate whether each of the following interpretations of grade equivalent (GE) scores are **appropriate (A)** or **inappropriate (I)** by circling the appropriate letter.

A I 1. A student who obtained a GE score of 3.1 in the spring of grade 4 would be expected to obtain a GE score of 4.1 in the spring of grade 5.

A I 2. A student in grade 4 who obtained a GE score of 4.7 in April scored higher than about half the students in the grade 4 norm group.

A I 3. A student with GE score of 5.3 in reading and 6.1 in math is performing better in math than in reading.

A I 4. A student who has GE scores in all subjects that are more than 1.5 above grade placement should probably be skipped to the next grade.

A I 5. A GE score of 11.0 for a grade 6 student indicates that the student did exceptionally well on the 6th grade content, but not that he or she could do grade 11 work.

A I 6. One year gains in GE scores for two students of 3.0 to 4.5 and 5.0 to 6.5 indicate equivalent amounts of progress.

OUTCOME: State the advantages and limitations of grade equivalent scores.

Directions: Briefly state some of the major advantages and major limitations of grade equivalent scores.

Advantages:

Limitations:

(Turn up bottom of page for answers)

Answers to SELECTION (17-C)

1.I 2.A 3.I

4.I 5.A 6.I

Exercise 17-D

RELATIONSHIP OF DIFFERENT SCORING SYSTEMS

OUTCOME: Converts scores from one scoring system to others.

Directions: Complete the following table by converting the given scores into comparable derived scores in the other scoring systems. Round your answers to the nearest whole number, except for z-scores where one decimal place should be reported. Assume that all score distributions are normal and based on a common reference group. The first row of scores, form a given z-score of 1.0 has been completed to illustrate the procedure. Try to complete the exercise without looking at the table in your book.

		Standard Age Score		
z-score	T-score	(SD=16)	Stanine	Percentile Rank
1.0	60	116	7	84
-1.0				
0.5				
-0.5				
	35			
	70			
		100		
		124		
0.7				
-1.3				

OUTCOME: Explains the value of using score bands on test profiles.

Directions: Explain why it is desirable to plot scores on a test profile as **score bands** instead of specific **score points**.

(Use Figure 17.2 in your book to check your converted scores.)

Exercise 17-E

INTERPRETATIONS OF SCORES ON PUBLISHED TESTS

OUTCOME: Distinguishes between appropriate and inappropriate interpretations of scores on published tests.

Directions: Indicate whether test specialists would **agree (A)** or **disagree (D)** with each of the following statements about interpretations of scores on published tests by circling the appropriate letter.

A D 1. Percentile ranks of tests of the same subject can be used interchangeably.

A D 2. The percentile rank score does not require an assumption of a normal distribution.

A D 3. Information about a student's previous educational experiences and language background is used in interpreting test scores.

A D 4. A band that extends one standard error of measurement above and below a student's observed score helps guard against overly precise interpretations.

A D 5. Because the units on a grade equivalent scale are approximately equal, the difference between 4.0 and 5.0 can be treated as equivalent to that between 8.0 and 9.0.

A D 6. Before making an important decision based on a test score, the interpretation should be verified by other evidence.

OUTCOME: States the advantages and limitations of using local norms.

Directions: Briefly state the advantages and limitations of using local norms to interpret test performance.

Advantages:

Limitations:

(Turn up bottom of page for answers)

Answers to SELECTION (17-E)

1.D 2.A 3.A

4.A 5.D 6.A

Exercise 18-A

ACCOUNTABILITY

OUTCOME: Identifies factors related to the press for accountability in education.

Directions: Indicate whether test specialists would **agree (A)** or **disagree (D)** with each of the following statements concerning the press for accountability in education by circling the appropriate letter.

A D 1. Pressures for accountability have resulted in increased school testing.

A D 2. Pressures to raise test scores have led to an increased emphasis on reasoning and problem solving.

A D 3. There is public support for the use of test results to compare schools.

A D 4. Test-based comparison of schools has definitely improved education.

A D 5. Concerns that accountability leads to teaching to the test have contributed to calls for increased reliance on performance-based assessments.

A D 6. The "Lake Wobegon" effect provides support for the claim that accountability has helped improve education.

OUTCOME: Lists the possible influences of accountability testing programs on the schools.

Directions: List the desirable and undesirable influences that high-stakes comparisons of schools on the basis of achievement test results have on the school.

Desirable:

Undesirable:

(Turn up bottom of page for answers)

Answers to SELECTION (18-A)

1.A 2.D 3.A

4.D 5.A 6.D

Exercise 18-B

NATIONAL AND INTERNATIONAL ASSESSMENT

OUTCOME: Identifies characteristics and limitations of national and international assessments.

Directions: Indicate whether the following statements about national and international assessment are **true (T)** or **false (F)** for circling the appropriate letter.

T F 1. The National Assessment of Educational Progress (NAEP) enables schools to compare the performance of their students to the nation as a whole.

T F 2. NAEP provides a means of monitoring trends in the achievement of students over more than two decades.

T F 3. In addition to national results, NAEP now provides results for state-by-state comparisons on a voluntary basis.

T F 4. NAEP collects achievement data for students by both age and grade level.

T F 5. Comparisons of nations based on international assessments are as trustworthy as comparison of regions of the country based on NAEP results.

T F 6. Differences in the selectivity of educational systems in different countries complicate international comparisons.

OUTCOME: Identifies influences at the national level that may influence the role and nature of testing and assessment in the future.

Directions: Briefly describe actions of the federal government that are likely to influence testing and assessment in the future.

(Turn up bottom of page for answers)

Answers to SELECTION (18-B)

1.F 2.T 3.T

4.T 5.F 6.T

Exercise 18-C

ASSESSMENT OF TEACHERS

OUTCOME: Identifies characteristics of existing teacher testing systems and current efforts to develop a system of teacher assessment.

Directions: Indicate whether each of the following statements about the testing and assessment of teacher is **true (T)** or **false (F)** by circling the appropriate letter.

T F 1. Passing the National Teacher Examination is required for national certification.

T F 2. In most states, all prospective teachers are required to pass the same subject-matter examination for certification.

T F 3. Teacher certification tests frequently include measures of basic skills in reading and arithmetic.

T F 4. Most teacher certification tests are justified on the basis of demonstrated predictive validity with performance in the classroom as the criterion.

T F 5. A number of efforts to expand assessments of teachers beyond paper-and-pencil tests have been initiated.

T F 6. It is expected that future teachers will have to pass the assessments constructed by the National Board of Professional Teaching Standards before they can teach.

OUTCOME: Lists desirable and undesirable characteristics of traditional paper-and-pencil tests required for teacher certification.

Directions: List some desirable and undesirable characteristics of paper-and-pencil teacher certification tests that have been widely used by states during the past decade.

Desirable:

Undesirable:

(Turn up bottom of page for answers)

Answers to SELECTION (18-C)

1.F 2.F 3.T

4.F 5.T 6.F

Exercise 18-D

CURRENT TRENDS IN EDUCATIONAL MEASUREMENT

OUTCOME: Identifies factors related to current trends in testing and assessment.

Directions: Indicate whether measurement specialists would **agree (A)** or **disagree (D)** with each of the following statements concerning current trends in testing and assessment by circling the appropriate letter.

A D 1. Computers are especially useful for adaptive testing.

A D 2. Computer-administered simulations of problems enable the measurement of complex skills not readily measured by paper-and-pencil tests.

A D 3. Despite concern about the quality of school programs, there has been a demand for **less** testing and assessment.

A D 4. Tasks requiring extended responses have been the target of most criticisms of testing and assessment.

A D 5. The focus on the consequences of testing and assessment has increased in recent years.

A D 6. The belief that testing and assessment shapes instruction has led to increased emphasis on performance assessments.

OUTCOME: Describes advantages and limitations of expanded uses of performance-based assessments.

Directions: Briefly describe some of the important advantages and major limitations of expanded uses of performance-based assessments.

Advantages:

Limitations:

(Turn up bottom of page for answers)

Answers to SELECTION (18-D)

1.A 2.A 3.D

4.D 5.A 6.A

Exercise 18-E

CONCERNS AND ISSUES IN TESTING AND ASSESSMENT

OUTCOME: Identifies factors related to concerns and issues in testing and assessment.

Directions: Indicate whether test specialists would **agree (A)** or **disagree (D)** with each of the following statements describing concerns and issues in testing and assessment by circling the appropriate letter.

A D 1. Many of the criticisms of testing are the result of misinterpretation and misuse of test scores.

A D 2. A common misinterpretation of scores on tests and assessments is to assume they measure more than they do.

A D 3. Testing does **not** have any undesirable effects on students.

A D 4. If a particular group receives lower scores on a test, it means the test is biased against members of that group.

A D 5. It is good practice to post scores on standardized tests so that students in a class can see how their performance compares to that of their peers.

A D 6. Test anxiety may lower the performance of some students.

OUTCOME: Lists the possible effects of students and parents examining school testing and assessment results.

Directions: List the advantages and disadvantages of the legal requirement that students and parents must be provided with access to school testing and assessment records.

Advantages:

Disadvantages:

(Turn up bottom of page for answers)

Answers to SELECTION (18-E)

1.A 2.A 3.D

4.D 5.D 6.A

Exercise A-1

MEASURES OF CENTRAL TENDENCY

OUTCOME: Distinguishes among measures of central tendency.

Directions: For each of the following statements, indicate which **measure of central tendency** is being used by circling the appropriate letter using the following key.

KEY: A = Mean B = Median C = Mode

A B C 1. It is the most frequent score in a set of scores.

A B C 2. It takes into account the numerical value of each score.

A B C 3. It is **always** an actual score.

A B C 4. It is **always** equal to the 50th percentile.

A B C 5. It is determined by dividing the sum of a set of scores by the number of scores.

A B C 6. It would **not** change if an extremely high score earned by a single individual was deleted from the set.

OUTCOME: Selects the measure of central tendency that is most appropriate for a particular use.

Directions: For each of the following statements, indicate whether the mean (A) or the median (B) would be most appropriate by circling the letter.

A B 1. To use with the quartile deviation.

A B 2. To use with the standard deviation.

A B 3. To divide a set of scores into two equal halves.

A B 4. To compute a set of standard scores.

A B 5. To limit the influence of a single score of 85 when all the other scores range between 35 and 65.

A B 6. To report the most widely used measure of central tendency.

(Turn up bottom of page for answers)

Answers to SELECTION (A-1)

(Top)	1. C	2. A	3. C	4. B	5. A	6. C
(Bottom)	1. B	2. A	3. B	4. A	5. B	6. A

Exercise A-2

MEASURES OF VARIABILITY

OUTCOME: Distinguishes among the measures of variability.

Directions: For each of the following statements, indicate which **measure of variability** is being described by circling the appropriate letter using the following key.

KEY: A = Standard deviation B = Quartile deviation C = Range

A B C 1. It is based on the highest and lowest scores only.

A B C 2. It is half the distance between the 25th and 75th percentiles.

A B C 3. It is also called the semi-interquartile range.

A B C 4. It takes into account the numerical value of each score.

A B C 5. It is influenced least by adding one extremely low score.

A B C 6. It can be used to identify the range of the middle 68 percent of scores in a normal distribution.

OUTCOME: Selects the measure of variability that is appropriate for a particular use.

Directions: For each of the following statements, indicate which measure of variability would be most appropriate by circling the letter using the following key.

KEY: A = Standard deviation B = Quartile deviation C = Range

A B C 1. To obtain the most stable measure of variability.

A B C 2. To obtain the simplest and quickest estimate of variability.

A B C 3. To obtain the range of the middle 50 percent of a set of scores.

A B C 4. To compute the amount of error in test scores.

A B C 5. To use with a small set of scores that includes one extremely high score.

A B C 6. To compute a set of T-scores.

(Turn up bottom of page for answers)

Answers to SELECTION (A-2)

(Top)	1. C	2. B	3. B	4. A	5. B	6. A
(Bottom)	1. A	2. C	3. B	4. A	5. B	6. A

Exercise A-3

CONSTRUCTING GRAPHS AND COMPUTING MEASURES OF CENTRAL TENDENCY AND VARIABILITY

OUTCOME: Constructs graphical representations of scores and computes measures of central tendency and variability.

Directions: Use the following set of scores to:(1) construct a frequency polygon with a class interval of three, (2) construct a stem and leaf diagram, and (3) compute the mean, median, range, and standard deviation.

Student	Score
A	60
B	58
C	56
D	54
E	53
F	52
G	48
H	47
I	45
J	42
K	40
L	39
M	37
N	35
O	32
P	29
Q	28
R	20
S	15
T	10

Median =
Range =
Mean =
Standard deviation =

(Turn up bottom of page for answers)

Answers to SELECTION (A-3)

STEM	LEAF
6	0
5	23468
4	02578
3	2579
2	089
1	05

Median = 41
Range = 50
Mean = 40
Standard deviation = 14

Exercise A-4

CORRELATION COEFFICIENT AND REGRESSION

OUTCOME: Identifies the characteristics of the product-moment correlation coefficient.

Directions: Indicate whether each of the following features describes the product-moment correlation coefficient by circling **Yes** (if it does) and **No** (if it does not).

Yes No 1. During the computation, it takes into account the numerical value of each score.

Yes No 2. Is easy to compute without the aid of a calculator.

Yes No 3. Can be used to compute estimates of test reliability.

Yes No 4. Can be used to evaluate test-criterion relationships.

Yes No 5. The degree of relationship is shown by the **plus** and **minus** signs.

Yes No 6. Can be used to indicate the cause/effect relations between measurement variables.

OUTCOME: Uses the regression equation to obtain predicted criterion scores from test scores.

Directions: The regression equation for predicting a criterion measure, Y, from a test score, X, is: predicted Y = -1.0 + .4X. Find the predicted criterion score for the following three students.

Carlos: Test score = 20

Predicted criterion score =

Kim: Test score = 15

Predicted criterion score =

Sam: Test score = 10

Predicted criterion score =

(Turn up bottom of page for answers)

Answers to SELECTION (A-4)

1. Y   2. N   3. Y   4. Y   5. N   6. N

Predicted criterion scores: Carlos, 7; Kim, 5; Sam, 3.

Exercise A-5

COMPUTING THE PRODUCT-MOMENT CORRELATION COEFFICIENT

OUTCOME: Computes the product-moment correlation coefficient.

Directions: Compute the product-moment correlation for the pairs of scores in the following table.

Student	X	Y	X^2	Y^2	XY
A	15	16			
B	18	15			
C	12	8			
D	13	11			
E	19	17			
F	10	9			
G	14	13			
H	11	5			
I	17	17			
J	11	9			

(Turn up bottom of page for answers)

Answers to SELECTION (A-5)
Product-moment correlation = .89